T0065021

Mushrooming Without Fear

It is easy to do without silver and gold
and your toga and cloak
but it is difficult to give up mushrooms.

Martial, Roman poet (AD 40–102)
Epigrams, Book 13, Number 48

Mushrooming Without Fear
The Beginner's Guide to Collecting Safe and Delicious Mushrooms

Alexander Schwab

Consultants Monika Lehmann and Roy Mantle

Skyhorse Publishing

Text and photographs © by Alexander Schwab, 2006

First published by Merlin Unwin Books Ltd (Ludlow, UK) in 2006

All Rights Reserved. No part of this book may be reproduced in any manner without the express written consent of the publisher, except in the case of brief excerpts in critical reviews or articles. All inquiries should be addressed to Skyhorse Publishing, 307 West 36th Street, 11th Floor, New York, NY 10018.

Skyhorse Publishing books may be purchased in bulk at special discounts for sales promotion, corporate gifts, fund-raising, or educational purposes. Special editions can also be created to specifications. For details, contact the Special Sales Department, Skyhorse Publishing, 307 West 36th Street, 11th Floor, New York, NY 10018 or info@skyhorsepublishing.com.

Skyhorse® and Skyhorse Publishing® are registered trademarks of Skyhorse Publishing, Inc.®, a Delaware corporation.
www.skyhorsepublishing.com

Library of Congress Cataloging-in-Publication Data

Schwab, Alexander.
 Mushrooming without fear : the beginner's guide to collecting safe and delicious mushrooms / by Alexander Schwab ; consultants, Monika Lehmann and Roy Mantle.
    p. cm.
 ISBN-13: 978-1-60239-160-4 (alk. paper)
 ISBN-10: 1-60239-160-2 (alk. paper)
 1. Mushroom culture. 2. Mushrooms. I. Lehmann, Monika. II. Mantle, Roy. III. Title.

SB353.S39 2007
635'.8—dc22

                         2007021372

20

Picture credits:
Frank Moser: cep page 40, middle picture cep page 42, orange birch bolete, page 68
Thomas Beer: larch boletes bottom right and middle right, page 57
Niels Sloth (www.biopix.dk): pestle shaped puffball, page 90

Every effort has been made to insure the accuracy of the information in this book. In no circumstances can the publisher or the author accept any liability for any loss, injury or damage of any kind resulting from an error in or omission from the information contained in this book.

Printed in China

Contents

Mushrooming Without Fear

It's Safe, Easy and Fun

Follow this guide and you can enjoy picking and eating delicious mushrooms uninhibited by fear and doubt. In order to enjoy a home-made gourmet mushroom dinner you do not really need to know what *basidia or hyphae* are. Nor do you need to know how to identify dozens of tasteless or poisonous mushrooms which you wouldn't want to eat in the first place.

Even advanced mushroom pickers face many uncertainties. The same mushroom looks different in the various stages of its development and different again when conditions are, for example, very wet or very dry. There are no end of edible mushrooms. Twenty-odd of these are of serious culinary interest, of which about a dozen are outstanding. Quality, not quantity, is what you are after. By good luck, the quality mushrooms are also among the most easily identifi-able mushrooms and some of these are among the most frequently encountered.

Mushrooms may look beautiful, but that does not mean that they are edible.

You wouldn't buy a soggy, wormy, half rotten mushroom in the supermarket. So why pick it in the wood? Quality is what the discerning mushroom hunter is after.

With the safe method explained in this book, you will be able to identify the best mushrooms positively by their unique and unmistakable features.

Your part in the safe method is:

1. As a beginner, leave alone most mushrooms you pass
2. Look closely at what you see before you, not at what you wish was there
3. Stick to the rules and tick off every stage in the mushroom identification process

The fly agaric is the most well-known species. For gnomes, pixies, fairies and other such folk it is safe. For us common mortals it is poisonous! Do not believe any old wives' tales regarding the identification of poisonous mushrooms. Stick to the safe method in order to identify the best edible mushrooms.

Why is the safe method safe? If you follow the safe method to the letter, you'll automatically steer clear of deadly mushrooms or mushrooms causing permanent damage.

In the families of mushrooms presented and recommended here, there is not one deadly mushroom.

This is the reason why the field mushroom is not in this guide.
The field mushroom belongs to the group of mushrooms with gills, and especially when young, the feld mushroom can be confused with some deadly species. The safe method therefore asks you to ignore all mushrooms with gills because it is within the gilled group that all the seriously poisonous and deadly mushrooms are found. If you're abso-lutely sure about the field mushroom, then pick it by all means. But

unless you are 100 percent certain, hunt for field mushrooms and other species with gills in the supermarket only.

Within the recommended groups you will find some slightly poisonous mushrooms but these happen to be so bitter, hot-tasting or strong-smelling that nobody sane would eat them anyway should they find their way into a meal (which they won't if you follow the safe method). Poisonous mushrooms in the recommended groups cause gastro-intestinal trouble of varying intensity. In other words: tummy ache. Nevertheless there is absolutely no reason for complacency: a harmless tummy ache together with an existing medical condition can be dangerous.

Now, enough with the warnings. With this guide in hand, you are sure to have a safe and memorable experience collecting delicious mushrooms. Good luck and have fun!

How to use this guide

1. Read the entire book **twice** in order to understand the approach

2. Study the "Gills," "Tubes," "Spines" and "Ridges" sections until you're confident you can correctly distinguish these features; then you're ready for a foray

3. Double-check at home by going through the full identification process step-by-step

The size of mushrooms

The size of mushrooms varies enormously, so any measurements given in this book are only a rough average. Don't be put off if you see an exception. The size can vary disproportionately due to weather conditions: if it is very dry, the mushrooms will be smaller than average. The other great variable is the growing medium. Don't be surprised if near a building site, for example, you find surprisingly large mushrooms growing out of sawdust, or where woodchips have been laid at the edges of paths.

What is a Mushroom?

A mushroom is a highly complex organism. Here are the essentials:

Fruitbody

The technical term for what is colloquially called the "mushroom" is the "fruitbody." Mycologically speaking, the real mushroom is the mycelium which is a fine structure of filaments on or in the forest soil. What we see as mushrooms (the fruitbody) are like the apples on an apple tree.

Cap

The caps of mushrooms have many different forms but what we're interested in is what is under the cap. Under the cap you'll find either . . .

Gills

Tubes

Spines

Ridges

Gills, tubes, spines and ridges produce, contain and release the spores. Wind and rain disperse the spores. Given the right conditions new mycelia and mushrooms form from the spores.

Stalk **Mycelium**

The mycelium

This is a close-up of about 4 mm of the mycelium above, as seen under a microscope. When conditions are right, the mycelium forms a fruitbody, pushing the mushroom up in order to disperse the spores or to wait for you to come along and pick it.

Rule number 1

Never, never take a mushroom with gills!!!

Gills

Identifying gills is the first and most important step. You must be sure that you're able to identify gills. Caps of gilled mushrooms come in many different shapes. Gills are the radiating blades on the underside of the cap. They fan out in a distinctly regular way. Gills have precise forms and come in many colors. Some of them are brittle some of them are soft. They can be rubbed off or separated from the underside of the cap quite easily. Gills are always attached to the stem or to the cap in a uniform way.

In perfect conditions the distinctly regular way in which the gills fan out is clearly visible.

Weather-beaten or old gills might be damaged or broken but on closer inspection their regularity will become apparent. Damage or no damage: either way they look as if drawn with a ruler.

In their regular way, gills differ in spacing and formation.
For example:

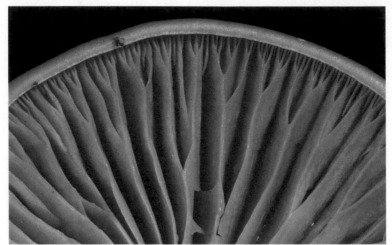

Gills can fork once or more.

Gills can be mixed: long and short gills.

Gills can be crowded.

Gills can be distant.

Gills are always uniformly attached to the cap or the stem. They start radiating from the same height along the stem or the same line around the cap.

Rule number 2

Only take mushrooms with tubes, spines and ridges and the mavericks portrayed in this book

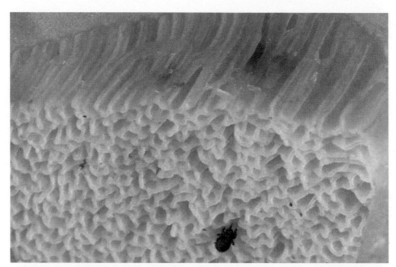

Tubes

Tubes are fine and tightly packed on the underside of the cap. The openings at the ends of the tubes are called pores.

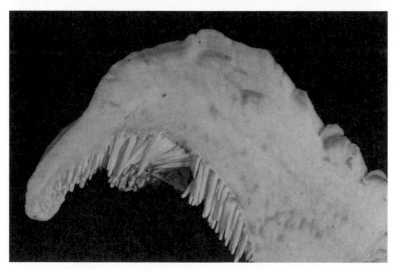

Spines

Spines hang like stalactites from the cap.

Ridges

Ridges are cross-veined, irregular and don't form a distinct pattern. Ridges are not attached to the cap: they are part of the cap.

Remember Rule number 1

Never, never take a mushroom with gills!!!

Tubes

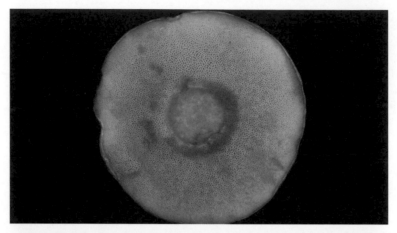

This is the underside of a cap with tubes. The tubes are fine and tightly packed and the openings at the end which are seen here are called pores. Tubes can be removed easily from the cap.

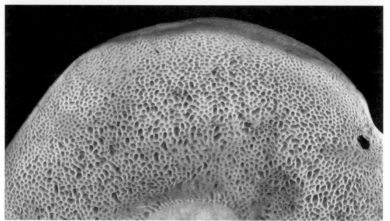

Pores form a regular pattern resembling a sponge.

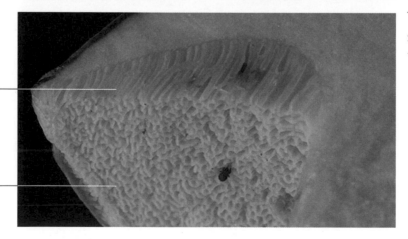

This cross-section shows the tubes and the pores.

Tubes

Pores

Spines

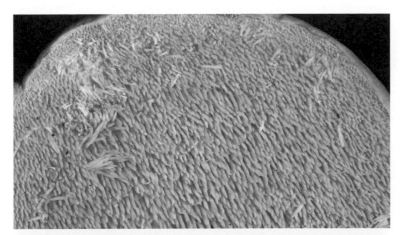

Spines hang like stalactites from the cap.

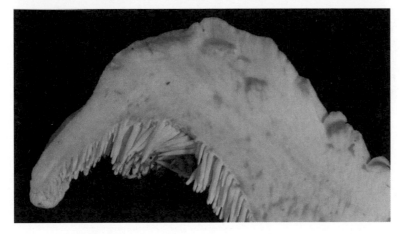

Ridges

Ridges are on the underside of the cap and are part of it. They are not just attached to the cap or stem: they're part of them, which is why you can't rub them off or pull them off easily.

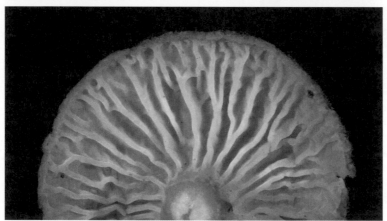

Ridges are cross-veined, irregular and they don't form a set, uniform pattern.

They are not attached to the stem in a distinct and regular way, as gills are: they are part of the stem.

The ridges are part of the stem. There is no distinct pattern in the way they grow out of the stem. Some ridges begin further up, some further down the stem. There is no regularity.

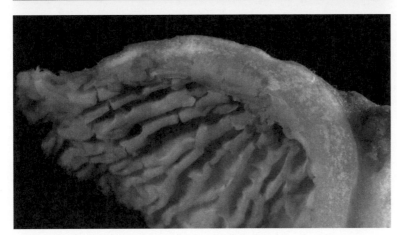

In contrast to gills and tubes, the section of a mushroom with ridges shows no typical features. Because of the irregular nature of ridges each section will look different whereas with tubes and gills you will find the same features each time.

How to Pick Mushrooms

Rule number 1 says not to take anything with gills. This also means you **do not touch or cut every mushroom you spot** in order to check for gills.

In nine out of ten cases you can see whether a mushroom has gills or not without even touching it. Looking under the cap by kneeling down provides good exercise.

If you really can't see what's underneath a cap, break off just a little to see what's there. If you want to be absolutely safe, you could use surgical gloves or a paper napkin which you discard later.

If you have established that you have, say, a mushroom with tubes and it is in reasonable condition and it tallies with the field identification card—then, and only then, should you pick it.

Rules:

Rule number 1

Never, never take a mushroom with gills!!!

This is our life insurance.

Rule number 2

Only take mushrooms with tubes, spines and ridges and the mavericks portrayed in this book

This means thoroughly understanding the information on pages 18–27 of this book.

Rule number 3

Only eat mushrooms which you have clearly identified with ALL the positive ID marks

The mushrooms you take must be a certain size in order to show all the identification marks. In their baby stage, so to speak, some deadly and poisonous mushrooms are almost indistinguishable from harmless species.

Rule number 4

If a mushroom smells rotten, it is rotten and if it feels soggy, it is soggy

There are a surprising number of people who believe that somehow a half decomposed, soggy, smelly mushroom will improve on the way home. It never does. Only take mushrooms which are in mint condition, that is, the flesh is firm and there are hardly any worms.

Rule number 5

Never, never eat wild mushrooms raw

The reasons for this are manifold but, to state the obvious, some areas in some woods are dog-ridden.

Rule number 6

Look before you cut

Always cut mushrooms, do not pull them out. You don't chop down an apple tree to pick the apples. Cut just above the ground and cover the remaining stem with a little mud or dead leaves. Whether this is of any conservational value is disputed but it does no harm. My recommendation is to cut and cover.

Picked mushrooms want to breathe. Do not suffocate them in plastic bags. Use a wicker or rattan basket or a linen bag. If possible use a proper mushroom knife with an integrated brush for cleaning. As the first cleaning of the mushroom should be done in the woods, the brush is very useful. However, any knife will do, as long as the blade can be folded into the knife handle, for safety. Common sense is your best adviser but be aware of any likelihood of ticks where you go mushrooming.

You might want a stick. It can be any kind of stick and you will find it helpful on the slopes. But you will also find it useful for pushing aside the grass and ferns, affording you a better look at the ground. Different mushrooms can be put together in your wicker basket . Better still, have a few paper bags in the basket; one paper bag (breathable) for each species. Putting, say, all trumpet chanterelles in the same bag helps you to spot the odd one out which you have inadvertently cut while not fully concentrating (see page 78).

Rule number 7

Mushrooms want to breathe:
Do not suffocate them in plastic bags

Either in the woods or at home.

Rule number 8

If in doubt, leave it

Now you're ready for a foray.

Positively Identifying Mushrooms

Always remember

Rule number 1
> Never, never take a mushroom with gills!!!

Rule number 2
> Only take mushrooms with tubes, spines and ridges and the mavericks portrayed in this book

Rule number 3
> Only eat mushrooms which you have clearly identified with ALL the positive ID marks

Rule number 4
> If a mushroom smells rotten, it is rotten and if it feels soggy, it is soggy

Rule number 5
> Never, never eat wild mushrooms raw

Rule number 6
> Look before you cut

Rule number 7
> Mushrooms want to breathe. Do not suffocate them in plastic bags

Rule number 8
> If in doubt, leave it

Mushrooms With Tubes

Cep or King Bolete

Boletus edulis

The cep is the king of mushrooms. Some truffles are rarer, pricier and confined to relatively small areas in France and Southern Europe. The cep on the other hand reigns supreme everywhere. Ceps can be found in the same location year after year. They can disappear in certain areas for a couple of years only to return spectacularly in masses. Some people gather only ceps which is understandable because of their beauty, taste and the thrill of finding them on the same spot time after time. Ceps vary greatly in general appearance, color and size. The size of the little fellow in the foreground of the picture above is about 2 inches in height whereas the giant is about 14 inches in height.

If conditions are right (starting off with a nice wet and humid spring or early summer) you should begin to look for ceps as early as the beginning of July. July and August are the months of the summer ceps.

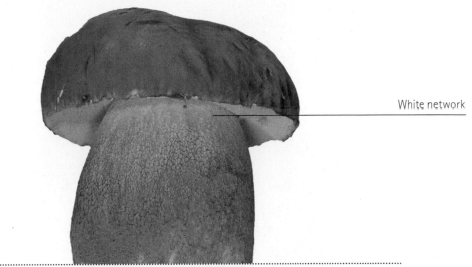

White network

White network

These are summer ceps. However all ceps, summer or autumn, show a fine white network on the top of the stem right underneath the tubes.

A classic autumn cep,
the king of mushrooms.

This is the key ID mark. The white net-pattern must be visible! All ceps, regardless of their time of emerging or stage of their development, have a fine white net-pattern at the top of their stalk.

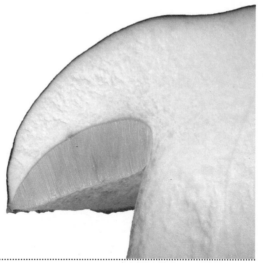

This is what a cep ideally looks like when cut. Not a single worm has even looked at this beauty. The tubes and the flesh do not change color when cut or touched.

Cep tubes are off-white and firm when young. Later they turn yellow-olive and are less firm. In maturity the tubes are olive. Tubes of mature ceps are soft as a sponge and the tubes of old ones are soggy and often like those of the big fellow on page 38. This does not always mean that the inside of old ceps is rotten but more often than not, it is.

Cep stalks vary greatly in appearance. These specimens clearly show the white net-pattern all over the stalk.

Nevertheless, it is that top section of the stalk just below the cap where it matters: there the white network must be clearly visible.

Positive ID Checklist
Cep or King Bolete

- [ ] Tubes, pores and flesh do not change color when cut or bruised
- [ ] Tubes are off-white, cream, yellow-olive or olive
- [ ] Pores do not show any pink tinge
- [ ] White network on top of stem
- [ ] Cap matches the color bar range (above)

Avg. size across cap:	5 inches when mature
Appearance:	July to November in eastern North America; July and August in the southern Rockies; September to November in the Pacific Northwest; November to February in California
Habitat:	Conifers, oak, beech, and birch in eastern North America; Engelmann spruce in the Rockies; pines and spruce in the west
Tip:	Where you find one cep, you'll often find others in the vicinity.

Red Cracked Bolete
Xerocomus chrysenteron

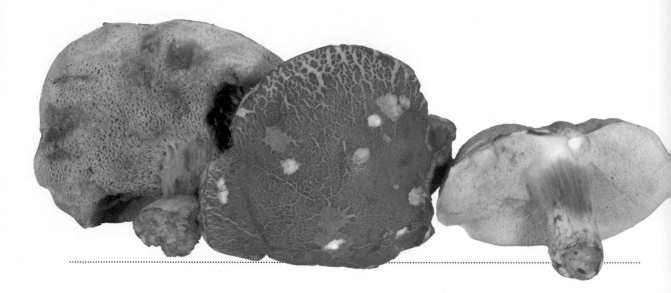

The red cracked bolete is probably the most common of all mushrooms with tubes. Its name describes its key identification feature perfectly. As the mushroom matures, the cracks become more prominent. Some people love eating the red cracked bolete, others quite like it and there are a few who don't take to it at all. The reason is its distinct fruity smell. In any case the red cracked bolete should only be picked as long as the pores are bright yellow and firm. The ideal size of the cap is about 1 inch across. Anything larger will make your dish all slimy.

Chrysenteron means 'with golden-yellow flesh.'

The color of the cap ranges from a velvety dark brown to a light brown.

Stems vary in color from yellow, to yellow with a little red to yellow flushed with red. Note the bright yellow pores. The stem and the pores bruise blue. The intensity of the bruising varies. See picture opposite page, far left.

This specimen is too old for the kitchen. The pores have lost their brightness. The dull yellow signals that it's no good anymore. When the pores are this color, the mushroom feels soft-to-soggy.

This is the perfect specimen. It has bright, firm golden-yellow pores and a nice firm cap about 1 inch across.

The caps are invariably cracked or lightly damaged. The typical identification mark of the red cracked bolete is the red showing through the cracks or the eaten-away patches.

When cut, the white-to-bright-yellow flesh turns blue. This can be very light blue and confined to certain patches.

Positive ID Checklist
Red Cracked Bolete

- ☐ **Yellow tubes**
- ☐ **Yellow pores bruise blue (varying intensity)**
- ☐ **Flesh bruises blue (varying intensity)**
- ☐ **Cracks and damaged patches in cap show a distinct red tinge**
- ☐ **Cap matches the color bar (above)**

Avg. size across cap: 2 inches when mature

Appearance: June to October in eastern North America; fall in the Pacific Northwest; November to January in California

Habitat: Associated with a wide variety of trees, including hardwoods (like oak) and conifers

Tip: Take only small specimens (cap no wider than 1 inch across)

Hen of the Woods
Grifola frondosa

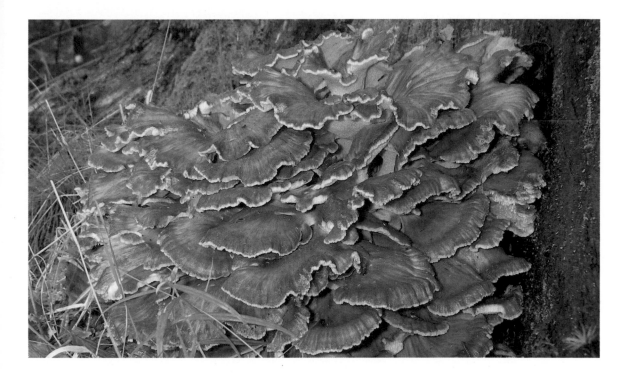

For thousands of years, the hen of the woods mushroom has been prized for its medicinal and culinary value in China and Japan. Maitake, the Japanese name for the hen of the woods, means "dancing mushroom" and one of the many interpretations of this name is that those who found a hen of the woods started to dance with joy because it was such a highly-prized mushroom.

It can be cultivated commercially. However, the wild hen of the woods mushroom you pick is infinitely superior because, firstly, you found it yourself; and secondly because it is as different from the farmed version as is a wild salmon to a farmed one. Finding a hen of the woods is indeed a reason to dance: a culinary event of the first order—and it's good for you too. There aren't that many things in life of which that can be said.

The hen of the woods is a cluster of fan-shaped overlapping caps.

On the underside of the cap (approx. 2 inches across) the tubes are at this stage visible to the naked eye. At this point the hen of the woods reaches gourmet status. Start picking now.

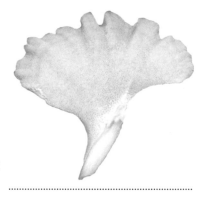

Actual size 2 inches

The fan-shaped caps with radial furrows are typical of the hen of the woods.

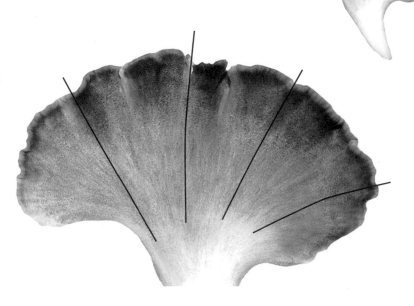

The tubes are clearly visible at this stage.

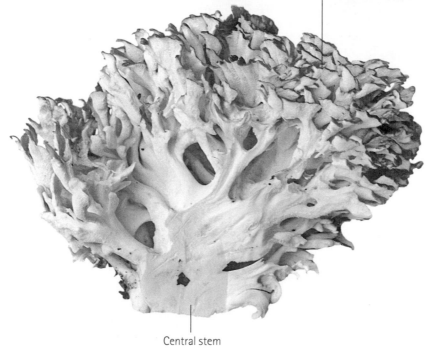

The cross-section shows a cauliflower-like structure with one central stem. The widest part of this particular specimen is approx. 12 inches.

Central stem

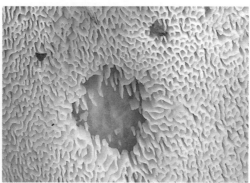

On older specimens the tubes are larger in diameter but remain whitish. As the tubes grow larger they almost look like spines—but tubes they are.

The main colors of the hen of the woods range from off-white, gray-beige, gray-brown, gray-black, brown-black to light brown-black. The hen of the woods can grow up to 30 inches across.

The culinary value of the hen of the woods depends on its age.

Size is not always an indicator of age. Some specimens grow very big, very fast. Despite their size these are still young specimens with white, firm, fibrous flesh with a pleasant smell.

Older specimens have a distinct unpleasant smell and their caps get more and more flabby and the edges crumble.

Positive ID Checklist
Hen of the Woods

- ☐ **Central stem**
- ☐ **Fan-shaped caps**
- ☐ **Caps with dark radial streaks**
- ☐ **White flesh does not change color when cut**
- ☐ **Cauliflower-structured cross-section**
- ☐ **White tubes**
- ☐ **Tubes do not change color when bruised**
- ☐ **Main cap color matches color bar (above)**

Avg. size across cap:	12 inches, but can grow up to 30 inches
Appearance:	Common to ubiquitous in the fall in eastern North America
Habitat:	Almost exclusively at the base of large oak trees
Tip:	Grows on the same spot for many years

Larch Bolete
Suillus grevillei

It is as if pure gold has grown out of the ground—and where there is one nugget, there are others. It's very rare to find a single larch bolete. More often than not larch boletes form a fairy ring. Always found near larch trees, this beautiful and delicious mushroom is covered with a yellow veil when young (above left). As the maturing mushroom grows, the veil breaks, leaving a transient ring on the stalk (above, center, and right). Always peel the cap of the larch bolete on the spot because otherwise it will make your basket and later your recipe all slimy. In wet conditions, the cap of the larch bolete is always slimy.

Grevillei refers to the Scottish mycologist R. K. Greville.

The veil still covers the cap in a 'baby' larch bolete.

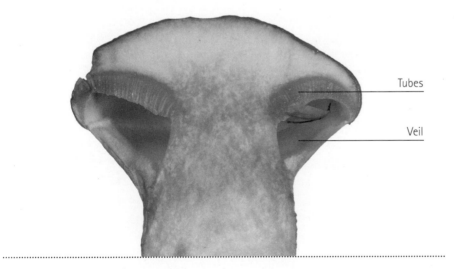

Tubes

Veil

The tubes must be clearly visible when you view a cross-section.

Tubes   Solid stem   Faint lilac
coloring   Veil

As the mushroom grows, the veil breaks. This picture (right) shows a larch bolete shortly after the breaking of the veil. The ring is transient (i.e., it often drops off) but even in the mature larch bolete you can always see a mark where the ring has been.

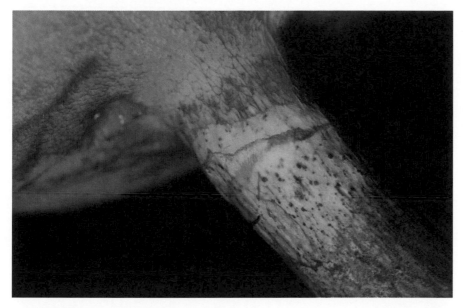

Above the ring marks: net pattern (can be very faint). Below the ring marks: mature specimens show distinct reddish-rusty streaks on the yellow stem.

If a larch tree is nearby and you see several gold nuggets, they are bound to be larch boletes. The beautifully golden (and slimy when wet) caps are the key sign. Positive identification however requires ticking off all of the identification marks.

Positive ID Checklist
Larch Bolete

- ☐ Yellow tubes
- ☐ Solid stem
- ☐ Bright orange-yellow or golden-yellow sticky cap (slimy when wet).
- ☐ Flesh in cap flushes lilac (faint)
- ☐ Found near larch trees
- ☐ Other specimens found in the vicinity
- ☐ Young: tubes covered with a veil
- ☐ Intermediate stage: ring visible, faint net-pattern above ring zone
- ☐ Mature: ring zone still visible, faint network above ring zone
- ☐ Cap color matches cap color bar (above)

Avg. size across cap:	2.5 inches when mature
Appearance:	Across northern North America
Habitat:	Always near larch
Tip:	Always remove the cap skin

Bay Bolete
Xerocomus badius

The bay bolete is an excellent mushroom and by good luck, it is also very common. The main bay bolete months are September and October but it can be found as early as June. If you find one, there are bound to be others nearby.

Badius means 'beautifully brown.'

Off-white pores:

This is the stage when the bay bolete is at its best. If the pores turn dirty yellow or green (see opposite page, top) the mushroom is past its culinary best. Take only specimens which have white or off-white pores.

Downwards-streaked stem:

The background color of the stalk is yellow-brown and is vertically frosted with brown streaks. The general appearance of the stalk ranges from light to dark brown but the vertical streaks will always be visible.

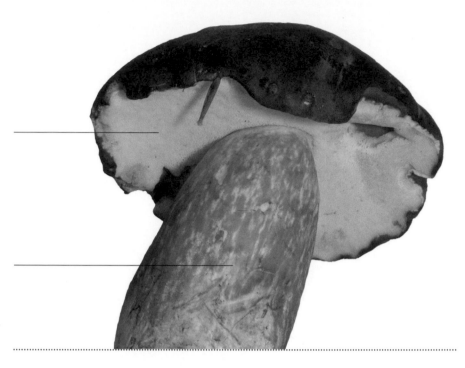

The off-white pores bruise green-blue.

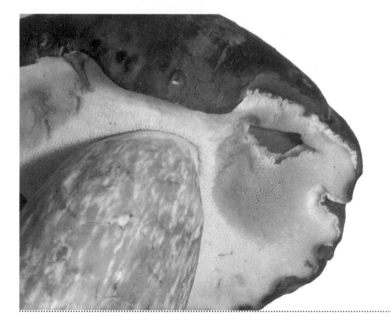

The change in color of the pores to green-blue varies in intensity. Here it's quite extreme.

The young bay bolete has pale yellow tubes which then turn to yellow, olive-yellow and dirty olive-yellow as the mushroom matures. Whatever the stage, the tubes and the flesh bruise blue.

The intensity of the coloring varies but pores, tubes and flesh will always bruise green-blue.

This is what the ideal bay bolete looks like: off-white pores and tubes, white to white-yellow flesh, without a trace of a worm.

Not as good: but if you discard the soft pores and cut away the wormy parts, it will still be excellent.

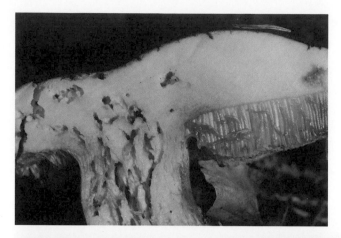

The color of the cap will always be "bay" (chestnut brown) but bay has a considerable range.

When wet, the cap
will be slightly slimy
and this will intensify
the color of the cap.

When dry, the color
of the cap will be
duller.

Positive ID Checklist
Bay Bolete

- ☐ Yellow tubes
- ☐ Off-white to gray pores (yellowish when aged) bruise green-blue
- ☐ Tubes and flesh bruise blue
- ☐ Stem is vertically frosted with brown streaks
- ☐ There is no net-pattern of any description on the stem
- ☐ Cap color matches cap color bar (above)

Avg. size across cap: 3.5 inches when mature

Appearance: June to November in northeastern North America; west to Minnesota

Habitat: Pine, spruce, and hemlock; often found growing on rotten logs

Tip: Remove soft tubes

Birch Bolete

Leccinum scabrum

The birch bolete is the most common of the "rough stalks." Rough stalks are mushrooms with a scaly stem. The birch bolete and its cousin, the orange birch bolete, are the most common rough stalks. Birch woods or groups of birch trees are the places to look for them. The birch bolete and the orange birch bolete have numerous relatives, all of which have a black or brown scaly stalk. They are all edible but it is recommended to stick to the two species presented here.

Scabrum means "rough."

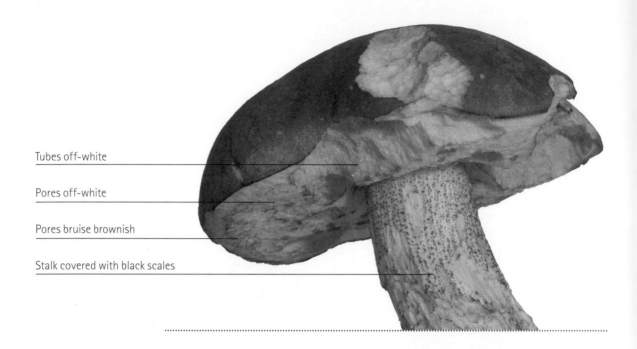

Tubes off-white

Pores off-white

Pores bruise brownish

Stalk covered with black scales

Close-up of stalk, showing black scales.

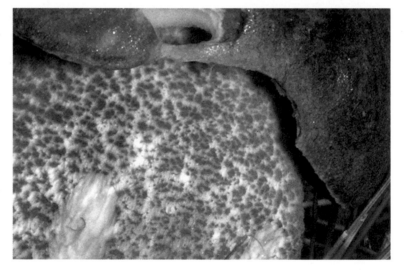

Pores and tubes change from off-white to a gray-white color as the mushroom matures. As a rule, you should only pick the birch bolete if it is firm and the pores are off-white.

Occasionally larger specimens are still nice and firm especially when conditions have been very dry.

These are the perfect birch boletes: a feast for the eyes and delicious to eat.

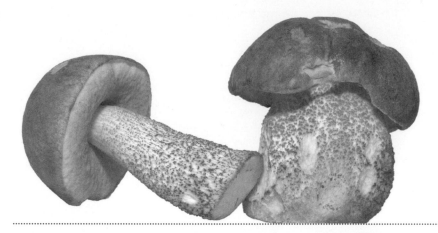

The orange birch bolete, *Leccinum versipelle*, is a beautiful mushroom. Its key identification mark is the stem covered with black scales, and of course the orange cap. When cut, the flesh turns faintly blue at the stalk base, displays traces of wine red and finally turns and stays gray. This color combination doesn't look especially appetizing but be assured the orange birch bolete is delicious.

Positive ID Checklist
Birch Bolete

- ☐ **Pores off-white**
- ☐ **Pores bruise brownish**
- ☐ **Stalk with small black scales**
- ☐ **Cap color matches cap color bar (above)**

Avg. size across cap:	3.5 inches when mature
Appearance:	July to October in eastern North America; winter in California
Habitat:	Birch; ornamental birch in California
Tip:	Discard stem and tubes of larger specimens

Always remember

Rule number 1

      Never, never take a mushroom with gills!!!

Rule number 2

      Only take mushrooms with tubes, spines, and ridges and the mavericks portrayed in this book

Rule number 3

      Only eat mushrooms which you have clearly identified with ALL the positive ID marks

Rule number 4

      If a mushroom smells rotten, it is rotten and if it feels soggy, it is soggy

Rule number 5

      Never, never eat wild mushrooms raw

Rule number 6

      Look before you cut

Rule number 7

      Mushrooms want to breathe. Do not suffocate them in plastic bags

Rule number 8

      If in doubt, leave it

Mushrooms With Ridges

Chanterelle
Cantharellus cibarius

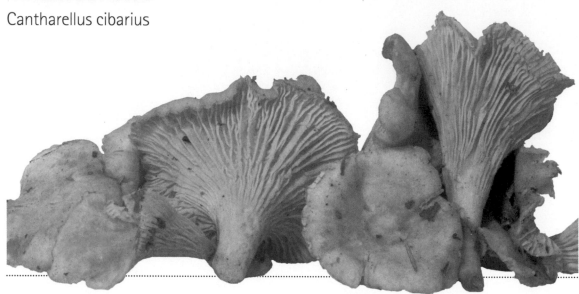

The chanterelle is not surprisingly the most popular mushroom.
It is delicious and very common. As it is imported throughout the year,
the supermarket is a good place to familiarize yourself with the
chanterelle. There is, however, nothing like the chanterelle you find
yourself. Refrain from picking tiny chanterelles, that is, little yellow
buttons which hardly show the identification marks. At some time
you'll come across one lonely chanterelle standing there in the middle
of nowhere for no particular reason at all. Leave it.

Cibarius means "belonging to food."

The classic color of the chanterelle is egg-yolk yellow which, however, can vary from light yellow or yellow-orange to yellow-ochre. The chanterelle does not change color when bruised. The entire mushroom is of the same color.

Young chanterelles have a flat cap with a slightly out-rolled margin. Do not pick any chanterelles smaller than 3 cm high.

The more mature the chanterelle, the more funnel-shaped it becomes.

The stalk is solid. The whitish to yellowish flesh does not change color when bruised or cut.

Whether young or mature, the chanterelle's ridges must be clearly visible.

Some mature forms can look highly irregular but on closer inspection, the funnel shape will still be there.

Positive ID Checklist
Chanterelle

- ☐ **Young specimen: Flat, wavy-edged cap, slightly out-rolled margin**
- ☐ **Mature specimen: Funnel-shaped**
- ☐ **Solid stem**
- ☐ **Found in groups on the ground, but not clustered at the base of trees; never on wood**
- ☐ **Ridges must be clearly visible**
- ☐ **Color matches cap color bar range (above)**

Be Careful:

Don't mistake the chanterelle for the Jack O'Lantern, a mushroom that grows across eastern North America. The Jack O'Lantern is orange, glows in the dark, has gills, and is not edible.

Avg. size across cap:	2 inches when mature
Appearance:	July to October in eastern North America; July and August in the southern Rockies; fall in the Pacific Northwest; November to February in California
Habitat:	Under conifers (pine and spruce) and hardwoods (especially oak)
Tip:	Watch for mossy patches with little plant cover

Trumpet Chanterelle
Cantharellus tubaeformis

In comparison to the attractively-colored chanterelle, the trumpet chanterelle (also known as the autumn chanterelle, and winter chanterelle) looks modest. On closer inspection, however, it is just as beautiful as the chanterelle. The trumpet chanterelle pops up overnight so quickly that you could watch it grow. It is an endearing mushroom not least because it tends to appear in great numbers and is frost-resistant.

Tubaeformis means "trumpet-shaped."

A cluster of trumpet chanterelles.

Color and shape vary according to age and weather conditions. Each individual trumpet chanterelle has more than one color. The cap is brownish and has a hole in the center. The ridges and the stalk range from brown-yellow to gray-yellow to gray-lilac.

The jagged edges here are due to frost.

The trumpet chanterelle has a hollow stem.

The advantage of having a paper bag for each species (see page 32) is seen here. You can spot the odd one out immediately.

Positive ID Checklist
Trumpet Chanterelle

- ☐ **Found in groups**
- ☐ **Brownish, thin-fleshed, wavy-edged cap with hole in center**
- ☐ **Hollow stem**
- ☐ **Ridges**
- ☐ **Cap color matches color bar (above)**

Avg. size across cap:	0.75–1 inch when mature
Appearance:	August to November in eastern North America; December to February in California
Habitat:	In conifer woods; on moss; on decayed, mossy logs
Tip:	Mossy banks are a favorite habitat

Always remember

Rule number 1

    Never, never take a mushroom with gills!!!

Rule number 2

    Only take mushrooms with tubes, spines, and ridges and the mavericks portrayed in this book

Rule number 3

    Only eat mushrooms which you have clearly identified with ALL the positive ID marks

Rule number 4

    If a mushroom smells rotten, it is rotten and if it feels soggy, it is soggy

Rule number 5

    Never, never eat wild mushrooms raw

Rule number 6

    Look before you cut

Rule number 7

    Mushrooms want to breathe. Do not suffocate them in plastic bags

Rule number 8

    If in doubt, leave it

Mushrooms With Spines

Hedgehog Fungus

Hydnum repandum

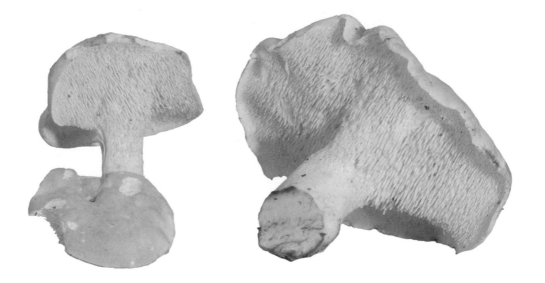

There are other mushrooms with spines but the only one of real culinary interest is the hedgehog fungus. The hedgehog fungus is a great delicacy. It grows on the ground but never on trees.

Repandum means "bent upwards" and refers to the way the cap often reveals its underside.

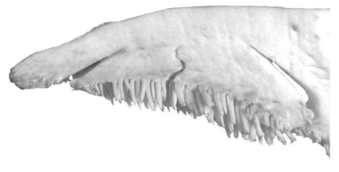

The spines are the key identification marks of the hedgehog fungus. The flesh is matte white and colors in places to a yellow-brown or rusty yellow.

Some hedgehog fungi grow in a very irregular fashion, up to the size of two fists. Nevertheless, all hedgehog fungi have distinctive spines. The specimens here show the entire color range of the cap. No other mushroom with spines has these colors.

Take only specimens where the spines are clearly visible. The hedgehog fungi pictured here are the perfect size for the kitchen. The cap size is 2.5–4 inches across.

Positive ID Checklist
Hedgehog Fungus

☐ **Spines**

☐ **Flesh matte white when freshly cut; flesh colors in places to yellow–brown or rusty yellow**

☐ **Cap margin wavy-edged**

☐ **Found on ground but not on trees**

☐ **Cap color matches color bar (above)**

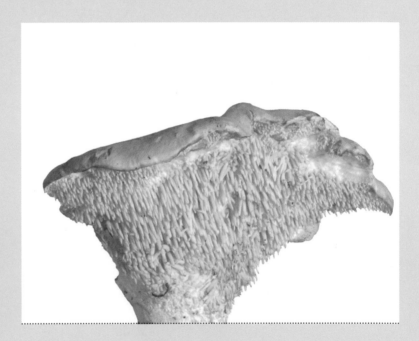

Avg. size across cap: 2.5 inches when mature

Appearance: August to November in eastern North America; October to November in the Pacific Northwest; December to February in California

Habitat: Common under oaks and mixed hardwoods in eastern North America; under conifers in the West and Pacific Northwest

Tip: Avoid older specimens; they are slightly bitter

Always remember

Rule number 1

    Never, never take a mushroom with gills!!!

Rule number 2

    Only take mushrooms with tubes, spines, and ridges
    and the mavericks portrayed in this book

Rule number 3

    Only eat mushrooms which you have clearly identified
    with ALL the positive ID marks

Rule number 4

    If a mushroom smells rotten, it is rotten and if it feels
    soggy, it is soggy

Rule number 5

    Never, never eat wild mushrooms raw

Rule number 6

    Look before you cut

Rule number 7

    Mushrooms want to breathe. Do not suffocate them in
    plastic bags

Rule number 8

    If in doubt, leave it

Mavericks

Common Puffball

Lycoperdon perlatum

The common puffball is a curious little mushroom. One cluster of common puffballs might number merely four specimens; the next about forty. The common puffball is a welcome addition to any mix of mushrooms. Cooked on its own, it is not everyone's cup of tea because of its distinct taste. The giant puffball is another matter and as good a reason as you'll get for a dinner party.

Perlatum means "widely spread."

The conical spines are visible to the naked eye and leave an unmistakeable net-pattern when rubbed off. Only pick specimens on which the white conical spines are clearly visible. The common puffball has no unpleasant smell.

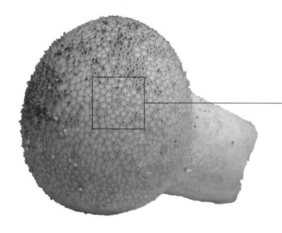

Net-pattern

These common puffballs are the perfect size (height 1–2 inches) and condition. When cut, the inside is all white and firm. Only when the flesh is uniformly white and firm, is the mushroom all right for use.

A variation of the common puffball is the pestle puffball. It too has spines but these are finer than those of the common puffball. It is pestle-shaped and it grows larger (height up to 8 inches) than the common puffball. It has no smell. If the flesh is uniformly white, and firm, it is good to eat.

The big brother of the common puffball is the giant puffball. Giant means it can grow to 3 feet in diameter. From a distance it looks all white but it can have a light yellow or yellow-brown tinge. The surface is smooth and feels rather like suede. The small specimen cut in two in the picture is about the size of a fist. When cut open, the flesh must be white and firm. Only then is the giant puffball edible—and delicious at that.

Positive ID Checklist
Common Puffball

- ☐ Conical spines
- ☐ Flesh must be firm and all-white
- ☐ Flesh undifferentiated, no thick rind or outline of mushroom visible in cross-section
- ☐ Net-patterned when spines are rubbed off
- ☐ Color white to off-white
- ☐ Found in groups
- ☐ No unpleasant smell
- ☐ Cap color matches cap color bar (above)

Be Careful:

Don't mistake the Puffball for the Stinkhorn, which has a thick rind and black-and-white interior, or the Destroying Angel, which has gills and a differentiated interior. They are not edible.

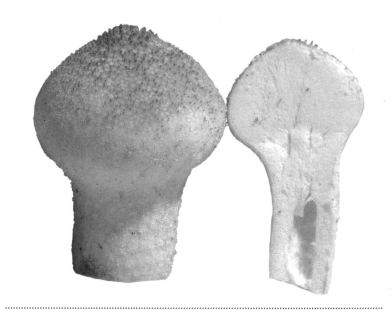

Avg. size across cap: Size of a golf ball. Giant puffball 12 inches

Appearance: July to November in eastern North America; fall in the Pacific Northwest; November to February in California

Habitat: Seems to feel at home everywhere, but especially in grassy areas

Horn of Plenty
Craterellus cornucopioides

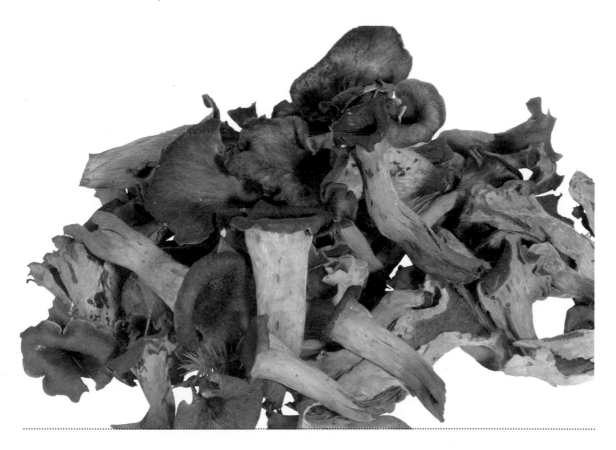

The horn of plenty is a real challenge to the mushroom forager.
Growing low on the forest-floor, it is so well-camouflaged that you can
look at masses of them and not see any at all. If you spot one, there
will be more: proceed with the utmost care because they are easily
trampled. The horn of plenty is without a shadow of doubt one of the
most delicious mushrooms in the world. But it has its price. Cleaning
the horn of plenty can be quite a job but the reward is well worth the
effort. Besides, when everybody joins the cleaning party, it is fun.

Cornucopioides refers to a cornucopia, or horn of plenty.

The horn of plenty is a funnel, the outer side of which looks smooth but in fact is slightly wrinkled. The color on the outer side is ash-gray, gray or pale gray with a bluish or lilac tint.

The inside of the funnel is brown, brown-gray, soot-gray or black. The surface is scaly or flaky and the top is out-rolled. In older specimens the top is wavy and split.

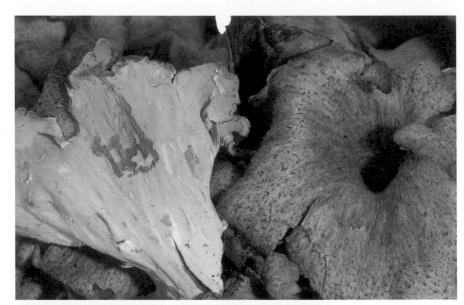

The horn of plenty seen from above in dry conditions.

The horn of plenty seen from above in wet conditions.

The funnel shape and
the out-rolled top are
typical.

This looks a rather
battered group of
horn of plenty. Some
of the tops have
already split. Don't be
put off—they still
taste good.

Positive ID Checklist
Horn of Plenty

- ☐ **Found in groups**
- ☐ **Funnel shaped**
- ☐ **Smooth-looking but slightly wrinkled outer surface**
- ☐ **Scaly inside surface**
- ☐ **Outer surface matches color bar above (top line)**
- ☐ **Inside surface matches color bar above (bottom line)**

Avg. size across cap: 0.75–1 inch when mature

Appearance: July to October in eastern North America; fall in the Pacific Northwest; January to March in California

Habitat: Under beech in eastern North America; conifers in the Pacific Northwest; oak in California

Tip: Always split them down the middle when cleaning and evict residents

Cauliflower Mushroom
Sparassis crispa

This is the mushroom of superlatives. For many people it is the very best of all. It may be a bit chewy but each bite releases a unique culinary sensation. If you fall for it, you'll be a dedicated cauliflower mushroom hunter forever. It resembles a cauliflower or a sponge. Once you learn to recognise its spicy smell, you could identify the cauliflower mushroom blindfolded. And then there is the size: it can grow so big that people simply overlook it. This specimen weighed in at 22.5 lb.

Crispa means "frizzy."

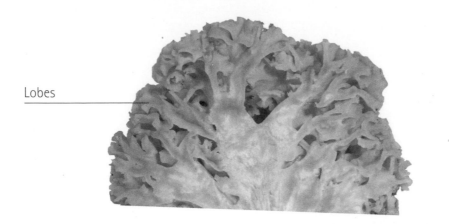

Lobes

Not all specimens grow to giant size. This one here is about the size of two fists. Its cauliflower or sponge-like appearance and the curved lobes are the key identification marks. There is nothing pointed or jagged in the cauliflower mushroom. The color ranges from creamy white to light brown. If it turns any browner than the specimen here, it has spoiled.

Curved lobes

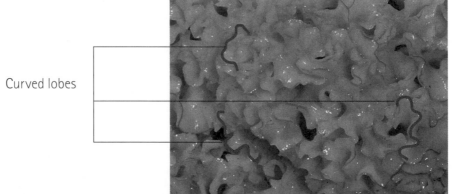

Positive ID Checklist
Cauliflower Mushroom

- [ ] **No gills, pores, tubes or spines**
- [ ] **No stem**
- [ ] **Looks like a cauliflower or sponge**
- [ ] **Lobes**
- [ ] **Matches range of color bar (above)**

Avg. size across cap:	12 inches but can grow up to three feet
Appearance:	July to October in eastern North America; fall in the Pacific Northwest; December to February in California
Habitat:	On pine stumps; at the base of oaks in the Northeast
Tip:	Grows on the same spot for many years

Mushrooming
Miscellany

The Mushrooming Regions

Eastern North America

Throughout this book, "Eastern North America" means regions primarily north of Georgia.

The Pacific Northwest

The Pacific Northwest consists primarily of Washington and Oregon (and parts of Idaho).

California

Throughout the book, "California" refers to the Bay Area, two hours north and south of San Francisco. Mushrooms also grow in Los Angeles and San Diego, but these appear toward the end of the California season.

As you can see, the timing and proliferation of mushrooms varies regionally. A good mushroom crop also depends heavily on weather conditions over the whole year. So the dedicated mushroom hunter need not confine himself to a mere couple of months of pleasure.

Trees and Mushrooms

Beech, birch, oak, pine, spruce, aspen and larch are the most important trees in mushroom-hunting, trees which are easily identifiable by their leaves or bark or both. It is most useful to know these five species of tree because if, for example, you explore a new wood, they will guide you to the mushrooms.

Beech

Especially:
horn of plenty

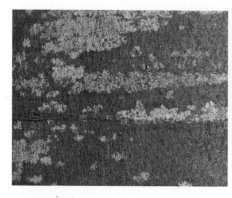

Oak

Especially:
chanterelle

Birch

Especially:
birch bolete

Pine

Especially:
cauliflower
mushroom

Larch

Especially:
larch bolete

The larch is the only
tree whose needles
change color in
autumn. You can't
miss it

Always remember

Rule number 1

> Never, never take a mushroom with gills!!!

Rule number 2

> Only take mushrooms with tubes, spines, and ridges
> and the mavericks portrayed in this book

Rule number 3

> Only eat mushrooms which you have clearly identified
> with ALL the positive ID marks

Rule number 4

> If a mushroom smells rotten, it is rotten and if it feels
> soggy, it is soggy

Rule number 5

> Never, never eat wild mushrooms raw

Rule number 6

> Look before you cut

Rule number 7

> Mushrooms want to breathe. Do not suffocate them in
> plastic bags

Rule number 8

> If in doubt, leave it

Handling,
Storage
and Cooking

Handling

The first rough cleaning of the mushroom should be done in the woods. Once you're at home you verify your spoils against the positive ID check lists and then—and only then—can you start the fine-cleaning. Do not wet mushrooms to clean them. Brush or wipe them with a cloth. Part of the cleaning process is to check the quality. Evict residents, cut away soft and soggy tubes and cut the remaining mushroom into bite-size pieces.

Storage

Drying

Drying is the classic way to store mushrooms. This method actually intensifies the taste of the mushrooms and they'll keep for years if you follow the procedure correctly. Do not dry anything you wouldn't eat fresh. Drying does not improve the overall quality of your mushrooms. There are various methods of drying mushrooms but there is only one way which guarantees consistent high-quality results and that is the only one I recommend: a dehydrator. Whatever the make, the principle is the same: warm air circulation. The sliced mushrooms are placed on trays and then dried in the dehydrator. Dried mushrooms are best stored in any air-tight container or a sealed plastic bag and then kept in a dark place. Mix all the species except the

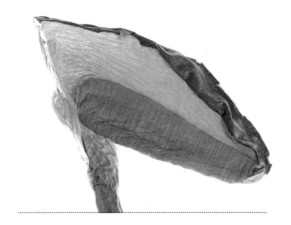

cauliflower mushroom. The more species in the mix, the better the taste. The cauliflower mushroom should be dried and stored separately (see page 115). **Do not dry chanterelles.** All other mushrooms in this book are suitable for drying. Dried mushrooms should be crackle-dry and snap when broken.

Freezing

Freezing is the other method. Results improved enormously when home vacuum-packaging machines became available. I recommend only freezing vacuum-packed mushrooms. Frozen mushrooms retain their color, texture and taste.

Never thaw frozen mushrooms!
Put deep frozen mushrooms in sizzling butter or hot water. Like this they keep their texture.

- Mix all the species except the cauliflower mushroom. Again: the more species in a mix, the better.
- I recommend deep freezing some cep, and horn of plenty, separately. You might want to add just a few of them for a particular dish or use the cep for a tasty starter.

The cauliflower mushroom should always be frozen separately (see page 115).

The chanterelle must be cooked before deep freezing – all other mushrooms in this book can be frozen raw.

Drying or freezing?

Either way, use **only perfectly fresh mushrooms.** Drying mushrooms
gives them that deep, strong mushroom taste. The frozen mushroom is
more subtle, but the colors of frozen mushrooms are as bright as on the
day they were picked and this adds considerably to the pleasure of
eating. Dried and frozen mushrooms can be mixed to get the best of
both worlds.

Cooking

Famous chefs win awards for stylish cuisine. Simplicity as a rule is not in their repertoire because no stars can be won by doing as little as possible. There are some really attractive mushroom books with some really attractive recipes but my advice is: do not try to emulate the recipes of the famous chefs – you'll never get there anyway. The place to enjoy grand cuisine is in the restaurant.

Do what you can do better than the professionals and do what they can't do and you're in for a feast. Besides, it's much less trouble, expense, and frustration to do it your own way rather than to follow recipes which pre-suppose the infrastructure of a five star establishment.

So where can't the professionals compete with you?

Chervil is a tasty and attractive herb for mushroom dishes

1. Freshness

This is not to say that restaurants use old mushrooms. They're just not as fresh as yours.

Your mushrooms have the shorter way from the wood to your kitchen. As a rule the restaurants' mushrooms come from a market and some of those mushrooms on the market have traveled a long way, some of them half way round the world.

2. Sourcing

You know where your mushrooms came from, whereas the chef in the restaurant, as a rule, knows only that it is the type of mushroom he wants. You know the very wood from which you picked them, and this is important because, for example, a cep can taste differently from one wood to another.

3. Quality

You know exactly where your mushrooms come from, and therefore you know their quality. In other words, you can be sure, for example, that your mushrooms grew in a non-polluted habitat or a dog-free zone.

Garnishing with finely chopped vegetables makes your mushroom dish a feast for the eyes.

4. Freedom

You are at liberty to vary the mushrooms that go into your dish and to change ingredients. In top restaurants, recipes are more precise and designed to be produced time and again. Your mushroom feast, on the other hand, can be unique each time.

5. Soul

There is a satisfaction in picking your own wild mushrooms and cooking them yourself that nothing can match.

Mushrooms are not garnish

The mushrooms in this book are of culinary value in their own right and should be treated accordingly. Treat a wild mushroom with the same culinary respect as you would a wild salmon.

Olive oil and mushrooms

Most contemporary mushroom cookbooks recommend olive oil for cooking mushrooms, but if you really want to draw out all the subtle flavors of your mushrooms and realize their full culinary potential, then olive oil is, generally, not a good idea. Unsalted butter brings out the best in your mushrooms.

Chives are the classic garnish for mushrooms.

Kills all known germs

There is a potential but very low risk of catching something nasty from a mushroom. To reduce that risk to zero, do not eat raw mushrooms and in the cooking process, increase the heat at one stage so that the dish either boils for a couple of seconds or sizzles in the butter. For that brief moment, turn the mushrooms in the pan so that they're exposed to the heat on all sides.

Curly or flat-leafed parsley underline the mushroom taste and add aesthetic appeal. Sprinkle on raw just before serving.

Special Cases

Chanterelle

The chanterelle must not be dried because it goes chewy. It must not be frozen without prior cooking because it will turn bitter.

Horn of Plenty

Some people like it on its own. Others use it only as a spice to add to a mix of mushrooms. In order to find out what you like best, dry and freeze separately.

Cauliflower mushroom

The cauliflower mushroom is also very distinct in taste and perhaps best on its own. Dry or deep freeze separately.

Basic preparation for fresh or frozen mushrooms

There's a lot of flexibility in these basic preparations. The amounts stated can vary considerably and you'll still get an excellent result. And the beauty of it is that you can't overdo mushrooms. The key to culinary success is the quality of the mushrooms.

Serves 4

What you need

Essential: 5–8 cups fresh or frozen mushrooms

4 cups water

1 cup white wine

1 teaspoon salt

1 shallot or small onion, chopped

1 cup cream

$1/4$ cup chives or parsley

Optional: A mix of finely chopped vegetables, such as carrot, zucchini, and celery: approx. $1/2$ cup

$1/4$ cup chervil

What you do

1. Put water, white wine, salt, and shallot into a pan

2. Do not thaw frozen mushrooms. Add frozen or fresh
   mushrooms and bring to the boil

3. Reduce heat and simmer until approx $^2/_3$ of liquid has evaporated

4. Add cream

5. Simmer again until approx. half the liquid has evaporated. Stir occasionally.
   If you feel there's too much liquid, simmer until it reduces still further.
   The less liquid, the more intense the taste and vice versa. Add water or cream
   to taste.

6. Add more salt and pepper to taste

7. You now have a wonderful mushroom sauce which you can serve with
   meat (chops, steak, etc.) or on its own with pasta or rice. Sprinkle with
   chives, parsley or chervil and the finely chopped vegetables

Basic preparation of dried mushrooms

Serves 4

What you need

Essential: 1–2 ounces dried mushrooms

4 cups lukewarm water

1 shallot or small onion, chopped

$^3/_4$ cup dry white wine

1 teaspoon salt

$^3/_4$ cup cream

$^1/_4$ cup chives or parsley

Optional: A mix of finely chopped vegetables e.g. carrot, zucchini, and celery: approx. $^1/_2$ cup

$^1/_4$ cup chervil

What you do

1. Take the dried mushrooms and put in a bowl or pitcher. Add 4 cups of lukewarm water. Soak for $1^1/_2$ hours (until the mushrooms float in brown liquid)
2. Put wine, shallot, salt, mushrooms and the brown liquid into a pan
3. Bring to the boil for 10 seconds Reduce heat and simmer until $^2/_3$ of the liquid has evaporated
4. Add the cream

5. Simmer until approx. half the liquid has evaporated. Stir occasionally. If there is too much liquid, simmer until further reduced. The less liquid the more intense the flavor, and vice versa
   Add water or cream to taste

6. Add salt to taste

7. You now have a wonderful mushroom sauce which you can serve with meat (chops, steak, etc.) or on its own with pasta or rice. Sprinkle with chives, parsley, or chervil and the finely chopped vegetables

Beyond the Stove

No summer or winter woodland walk will ever be the same again once you become aware of the amazing universe of the forest floor, which in terms of color and life is a match for any coral reef. And the forest floor is right at your feet, so to speak, or at least not far from your door.

When in autumn you walk through a beech wood and marvel at the intensity and variety of the colors, think for a moment about those leaves.

There are about 2 tons of leaves to one acre of beech wood. After their job is done on the branches, they fall to

the ground. If they did not decompose we would soon be swimming through a sea of leaves. Likewise, unrotted fallen branches would in time pile up as insurmountable obstacles.

A leaf falls to the ground, inspiring on its way a poet who happens to be passing by: but on a more practical level, bacteria are the first organisms to get busy.

Shortly afterwards, it is fungi, algae and protozoans which play their part in the work of decomposition.

Soon, millions of organisms and little animals are at work on that single leaf. In due course the leaf is fully decomposed, its constituent parts recycled back into the forest soil, keeping it fertile by providing the essential substances for all the plants in the wood.

The vital role of mushrooms

Fungi are key players in maintaining tree growth and health. The reason the birch bolete is always found near birch trees, and the larch bolete always near larch trees, is that these pairs live in a symbiotic relationship. They need each other.

The mycelium of the birch bolete and the fine root-ends of the birch tree link up and exchange substances. There is a lot of chemistry going on, with a net result of mutual benefit. There are intricacies and complexities in all matters fungi and there are, so to speak, vast uncharted areas on the fungal map. However, even the most half-hearted mushroom-hunter will soon become aware of the all-encompassing interrelations in the wood. Mushrooming makes you see and appreciate the woodland differently.

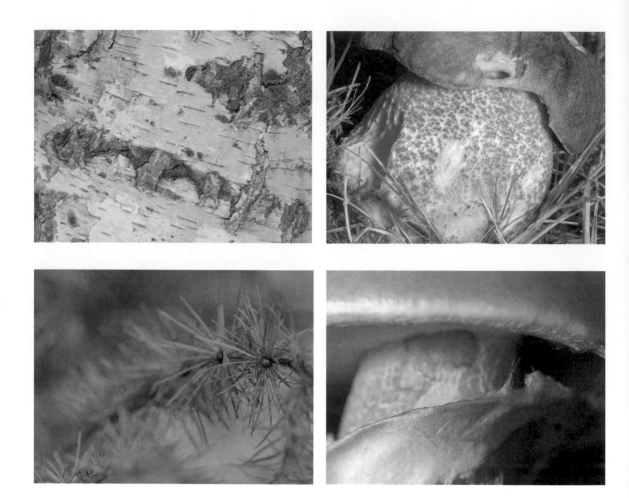

There is a practical point to these impressions which will help you in your mushrooming.

The similarity of patterning between the trunk of the birch tree and the stalk of the birch bolete is there for all to see. The similarity in color between the larch needles (once they change color in autumn) and the larch bolete seems obvious. These pointers can all act as nature's clues and reminders for the mushroom forager. But why is this so? Whatever the scientific answer, if indeed there is one, this similarity illustrates beautifully the interconnected nature or, if you prefer it, the mutual dependence, of life in the wood.

Perhaps even more intriguing is the similarity between the pine bark and the cauliflower mushroom, which grows near pine trees. It could be a coincidence or it could be a complex pattern at work. Either way, the cauliflower mushroom is utterly delicious – but if you mull over a conundrum like this, doesn't it add interest and pleasure to mushroom hunting?

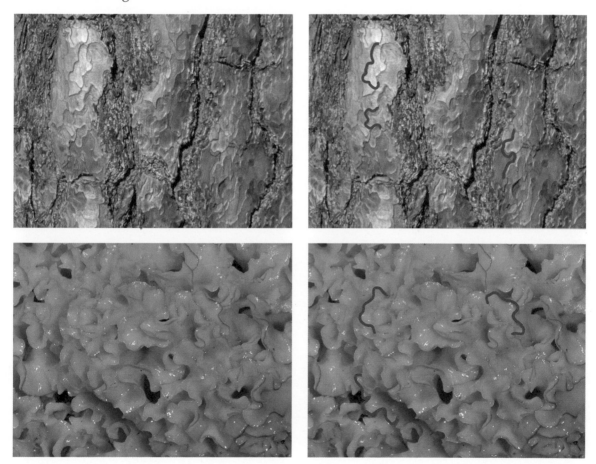

But I still want more advice on the secrets of successful mushrooming.

Mushroom hunting and the world of fungi have more than their fair share of mysterious lore, imponderables and surprises. Take that inexplicable instinct for the right spot to find mushrooms, for example. The experienced mushroomer knows his favorite haunts and whenever conditions are right, the mushrooms seem to call him. The dedicated mushroom hunter doesn't have to make a conscious effort to memorize the perfect weather conditions for a particular species. Nor does he have to keep a log on the wind direction. Over the years, he just knows that on the right day he has to go to spot X. As vague as that sounds, the mushrooming "instinct" may well have a rational explanation, but it needn't be a conscious one.

No exact science to mushrooming-hunting, then?

True, if you meet gnomes, pixies, fairies, elves or goblins in the woods, your rationality will be doubted. But when fingers of fog creep eerily through the lonely wood, mythical creatures don't seem far away. When out mushrooming in the twilight, you can easily understand how our forebears believed there were strange goings-on in the woods. Perhaps there

EIN GLÜCKLICHES
NEUES JAHR

were: and perhaps our modern powers of perception are just that much weaker than they used to be. The cold light of reason may have evicted all those colorful beings from the woods and now nobody is quite sure where they are anymore.

But they did leave their mark. The last bastion of mysticism is the fly agaric (left) which to this day has the power to fire the imagination. No decent fairy tale illustration is complete without the fly agaric. If the cep is the king of mushrooms, the fly agaric must be the queen. And as always with mushrooms, the mystical mixes with the scientific.

Fly agarics are useful indicators: where they are, there are often ceps nearby. Fly agarics wave and shout "Over here! Over here there are beautiful ceps!" The king is always near the queen.

Yet another mingling of fact and fiction: the fly agaric got its name from its surprising powers: it is a natural insecticide. Pieces of it are put in milk to attract flies. The flies then get drugged and die. In fact, the hallucinogenic properties of the fly agaric have inspired some mind-boggling theories and experiences.

Shamans, or holy priests, reportedly used the hallucinogenic fly agaric to make contact with the gods, while lesser mortals joined in the ritual by drinking the urine of the shaman. This would have been little more than water with the hallucinogenic substances still active in it, because, before the ceremony, the shaman had to fast for several days before ingesting the fly agaric.

The phrase "to go berserk" refers to the Berserkers (Viking warriors) who went fearlessly into battle and were reported to fight "like mad dogs and wolves." This is said to be the fly agaric's work.

It surely now comes as no surprise that Santa Claus is also linked to the fly agaric. After all, there are the same distinctive colors: and there he is in his reindeer-powered sleigh, flying all over the world, filling stockings, and involuntarily cleaning chimneys. Only a full-blown shamanic fly agaric ritual could take Santa Claus and come up with a tale like that!

These are just a few of the theories and ideas linked to the fly agaric. Fly agarics are also supposed to bring good luck, and in certain parts of the world they remain a popular emblem in advertising, on stamps, and on Christmas and New Year's cards. Luck is a fitting theme with which to conclude this book. Don't trust your luck when identifying mushrooms. Follow the instructions in this book and enjoy mushrooming without fear.